	DATE DUE		

LET'S GO TO A CONCERT

LET'S GO TO A
CONCERT

WRITTEN BY
LAURA SOOTIN

ILLUSTRATED BY
ROBERT EGGERS

G. P. PUTNAM'S SONS   NEW YORK

The author and artist wish to thank the following for their help in the preparation of this book: Otto Deri, Professor of Music, City College, New York; Mac Rudolf, Music Director, Cincinnati Symphony Orchestra, Cincinnati, Ohio.

© 1960 by Laura Sootin   Illustrations © 1960 by Robert Eggers
MANUFACTURED IN THE UNITED STATES OF AMERICA
Library of Congress Catalog Card Number: 60-5643
Published simultaneously in the Dominion of Canada by
Longmans, Green and Company, Toronto

During the week before a concert, the symphony orchestra works on the music for the performance. The musicians practice at home. They meet for rehearsals to work with the conductor. The orchestra members work together to make their playing better and better, so that they will play their very best at the concert. By visiting the concert hall during a rehearsal, you will find out how the orchestra prepares for a concert.

The concert hall is dark when you come to a morning rehearsal. Only the stage is lighted. On the stage you see many folding chairs, seats for the city's symphony orchestra. Next to the chairs are music stands to hold the music. A shiny golden harp is at the far right, toward the front of the stage. As you look, the musicians walk on in twos and threes, coming through a door at the back of the stage. They carry their instruments, and stroll in talking to one another.

The stage begins to fill up with musicians wearing slacks and sweaters or business suits. Many of them take out music sheets. They begin to practice the hardest sections of their parts for the next concert. Since each musician works on the part that is hardest, the players all practice different music. Strange combinations of sounds and music come from the stage.

Then the conductor arrives. He walks to the front of the stage and sits on a stool before a music stand, facing the orchestra. The musicians stop playing and talking and wait for him to speak.

The oboe sounds the A-note and the players tune their instruments. The conductor tells the orchestra that they will first rehearse the second movement of the symphony for the next concert. Everyone flips pages to find the right place. The conductor looks at the score or large sheets of music in front of him. He is the only one on stage who has a complete score, with parts for all the instruments shown on different lines of each page. From top to bottom, his score has lines for parts played by woodwind, brass, percussion, and string instruments. The players have parts only for their own instruments.

Now everyone has found his place in the music. The conductor begins to move his stick or "baton." At a downward movement of the baton, the orchestra begins to play. You see the violin bows moving smoothly up and down together, the musicians standing beside the drums counting beats until their part comes up, and then . . . the conductor gives a signal and all music stops.

"Let's go back to the beginning of the movement," he says. "Just the woodwinds play." He raises his baton and they begin. You hear only the flutes, oboes, English horn, clarinets, and bassoons. After a few notes, the conductor stops again. "Could we have it louder and tapering off?" He sings to show them what he means.

The woodwinds try it again, playing on for about five minutes. This time the conductor is satisfied. "Let's put it together. . . . Everybody!" The orchestra starts to play the movement again, and this time the conductor leads them through it without stopping. While the orchestra plays, the conductor's right hand, holding the baton, moves through the air with firm strokes. He points sharply at the horns when it is time for them to come in loudly.

The orchestra goes on to finish the symphony. When necessary the conductor interrupts to tell the players how he thinks the music should sound. After the symphony has

been rehearsed, the musicians stop for a fifteen-minute rest. Most of the players stroll out to the musicians' lounge in the concert hall. This room, furnished with chairs, tables, and coat racks, is a place to relax and to meet during free time at concerts and rehearsals.

When the musicians return to the stage, you hear more of the program for the next concert. The orchestra plays straight through some of the music, although the conductor stops a few times to go over and over difficult sections. Altogether, the morning rehearsal lasts three hours. Before each concert there are at least ten hours of rehearsal.

During the rehearsal, you saw the conductor at work as audience and judge of the music. He corrected wrong notes, told the orchestra how he wanted the music to sound, and gave advice on fingering and lip movements to help players make the music sound as he thought it should.

It's not hard to see how the conductor corrects mistakes. But his art is in the way he brings the music to life. He sees that the feeling of the music comes through to the audience by all the small corrections he gives at rehearsals, asking the orchestra to make certain parts faster, softer, sharper, sweeter, with sound evenly balanced between two sections of the orchestra or with one section stronger than the others. He also asks the players to stress certain notes, or to make the rhythm clearer to the audience.

If you think about what you have seen, you

will realize that the conductor began preparing for the concert long before the rehearsal. At home he thought about how each work of music should sound. To decide just how a piece should be played, he considered the composer's plan. He thought about other music by the same composer, and music by other composers who wrote at the same time in history. The conductor also considered how the shape and construction of the concert hall might change the sound of the music.

He went through his score, adding marks to the music and writing down playing instructions for the orchestra members. His changes and directions were planned to make the music sound best when played in the particular concert hall by his orchestra. He went on to mark the parts that would be most difficult for the orchestra to play. He also marked the parts that would be most important in giving the audience the feeling of the music. At the rehearsal, he watched for all the places he had marked and went over them thoroughly.

The conductor plans rehearsal time very carefully. He must be sure the whole concert program is ready after a few rehearsals. At each rehearsal, he plans to work on the most difficult parts first, while the musicians are still fresh.

Rehearsal is only part of the musicians' job in preparing for a concert. At home the sixty to one hundred musicians of an orchestra practice their parts for the concert program. They work on fingering and rhythm. They practice difficult parts of the score. They learn their parts well enough so that they can watch the conductor's signals during the rehearsals.

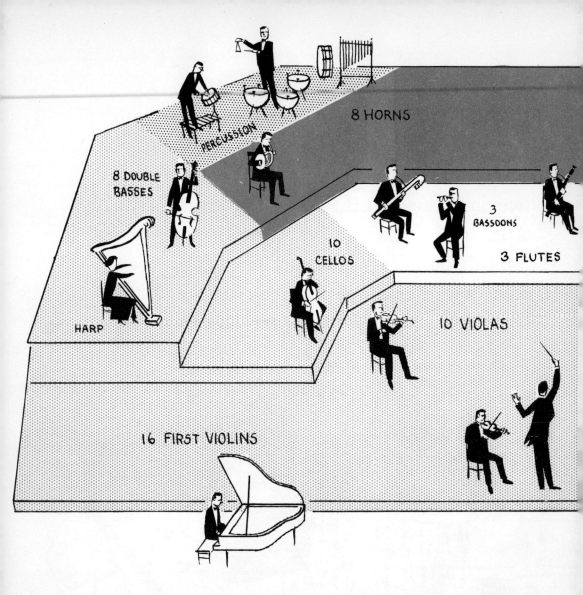

There are four families of musical instruments — and four main sections — in a sym-

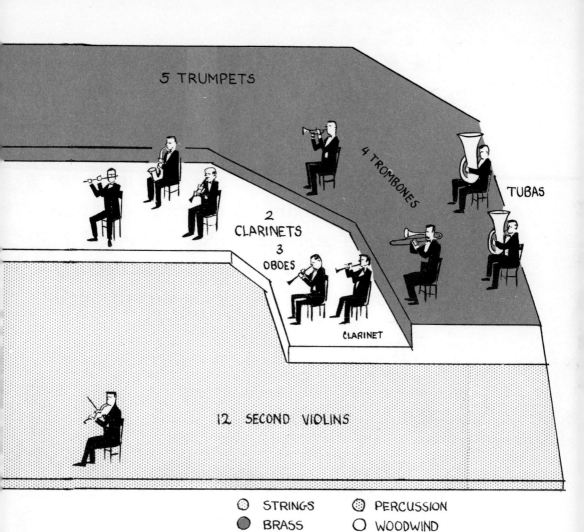

phony orchestra. They are the string, woodwind, brass, and percussion instruments.

The string instruments, which make up more than half the orchestra, are placed nearest the front of the stage. The violins are divided into two groups — the first and second violin sections. Usually the first violins are at the left of the conductor, closest to the front of the stage. The second violins are behind the first violin section. The violas, somewhat larger and sadder-toned strings, are slightly left in

cello

viola

bass

front of the conductor as he faces the orchestra. The cellos, strings whose sound is rich and mellow, are to the right of the conductor, about as far from the front of the stage as the second violins. The huge basses, deepest-voiced string instruments, are to the back and right of the cellos.

Behind the string sections of the orchestra you see the rows of woodwind instruments. The light-voiced flutes and piping piccolo are in front. Then come the oboes, English horn, mellow clarinets, and sad-voiced bassoons.

The back rows are for the shiny brass instruments. These instruments are always placed far back because their clear, strong voices are easily heard. From behind the strings and woodwinds they sound clearly, without drowning out the notes of weaker instruments. In the brass family are the warm-voiced, curving French horn, the bright-voiced trumpet, and the smooth-sounding trombone. The big toot of the brass section comes from the large mouth of the tuba.

trombone

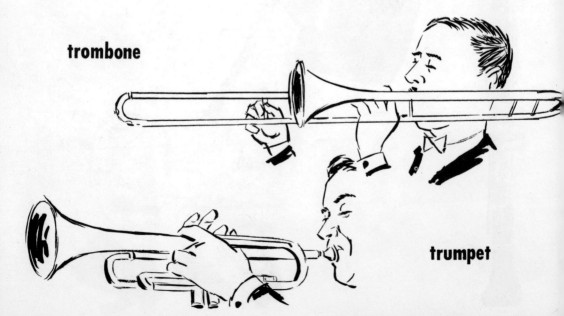

trumpet

To one side near the brasses you see the percussion section of the orchestra. Here are the big pots of kettledrums, the huge bass drum, the rattling side-drum. Here also are the clanging cymbals, the gong, bells, triangle, tambourine, and a few other percussion instruments. You can see that the percussion section has

fewer players than instruments. Each musician can play several instruments because percussion is not needed at all times in most concert music. During some works, however, the percussion players must leap from one instrument to another to keep up with the percussion score.

Orchestra sections may be seated in other ways besides this. The players may be shifted for certain works of music or because the conductor likes a different arrangement of sections. Some conductors prefer to have the first violins to the left and the second violin section

at the right front of the stage. And if a piano solo is part of the program, the piano is wheeled on and placed before the conductor, in front of the violins. You will never have trouble finding the large golden harp, whether it is at the far right front of the stage or out to the left behind the violins.

Each group of instruments has a leader. He is the orchestra's best player of his instrument. He plays alone whenever the music has a short solo part for his instrument. The leader of a section usually sits closer to the conductor than any other member of his section. He is placed so that the audience can hear him most clearly when he plays alone. A particularly important section leader is the leader of the first violin section. He even has a special title, that of "concertmaster." The leader of the percussion section plays the kettledrums. He is known as the "tympanist."

You may wonder what training the orchestra members have received, and how they went about joining the symphony orchestra. If you ask a musician during the rehearsal break, he will tell you that he began to study his instrument when he was in high school or even before that. Then he went on to study at a special music school called a conservatory of music. There he had more lessons in playing his instrument and also had a chance to play in the conservatory orchestra. In addition to training in the best methods of playing his instrument, he learned about the history of music and the different types of music. The young musician learned to play works of

music at first sight of the score. This skill of sight reading is necessary to every member of a symphony orchestra.

After years of schooling and home practice, the player may be ready to try out for a position in an orchestra. The test he takes is called an "audition." Judging the player at his audition may be the conductor alone, or a com-

mittee made up of the conductor and leaders of the different sections of the orchestra.

At the audition, the musician answers questions about his training and experience. He plays a piece of music he has prepared. The third part of his audition is a sight-reading test. For this test the judges have chosen several short passages from different works of music often played by symphony orchestras.

A conductor's training at a conservatory of music is much like that of an orchestra member. He must sight read very well, and know the history and all the different types of music. In addition, he must study the rules by which music is composed in order to recognize printed mistakes and also to change some scores to suit the instruments of his orchestra. He studies conducting, too, and practices by leading the conservatory orchestra. He must be an expert player of one string instrument and one wind instrument, and he must understand

the musical possibilities and difficulties of every instrument in the orchestra. He must also play the piano. When the young conductor leaves the conservatory, he looks for a job as assistant conductor of an orchestra.

The conductor and musicians are the orchestra people you see when you come to a concert. But there are many other people who do important work for the orchestra. Every

symphony orchestra has a librarian, who is in charge of all the scores. At the beginning of the concert season — which usually lasts seven months from fall to spring — the librarian orders the scores that will be needed for all concerts. Then, when the conductor marks changes or playing directions on his score, the librarian copies them on the parts for the musicians.

The orchestra also has a business staff to take care of concert arrangements. Working closely with the orchestra are people from the community. They raise money for the orchestra, and see that everyone in the city knows about the concerts.

There is a committee of community supporters. This group is called the orchestra association. It has elected officers and a board of directors.

A whole season of concerts is planned to present many types of music to the public, written at different times in history, by different composers, and in different styles. And each concert of the season follows a special plan in itself.

A single concert lasts about two hours. It may include the works of just one great composer. It may present works written at the same period in history by composers from dif-

ferent countries. It may be a concert of works by composers of one nationality, all written at different periods. Or the concert program might be chosen to present contrasting styles, periods, and types of music. When you get your ticket and read the program notice, you will know what to expect at the concert. You will discover what plan the concert follows, and you will see which works have been chosen to carry out that plan.

When you come to the concert, you will receive a program booklet to read before the music begins. In this program, you will see the names of all the orchestra players, usually listed by instrument. You will find program notes that give interesting information about the works of music to be played. At some concerts the conductor gives the program notes himself, talking to the audience about each work of music before it is performed.

Five minutes before the concert begins, the musicians file onto the stage, wearing dark suits and white shirts. The women members of the orchestra wear long dresses. The musicians sit with their backs very straight — good posture is important for fine playing. You hear strange and exciting sounds as each section tunes its instruments. The concertmaster enters. He leads the tuning of the orchestra.

When the conductor looks from the corner of the stage and sees the audience has arrived and the doors are closed, he walks slowly out and bows to the audience. At his special stand, called a "podium," he waits a moment for silence, raises his baton — and the concert begins.

You listen and forget yourself as beautiful music fills the hall. Your ears pick up the notes

from each section of instruments. You watch the leader of a section as he plays a short solo. At the end of the first work, you are surprised for a second by the silence. Then you come back from the world of sound and clap with the rest of the audience.

The concert goes on. Just before the intermission recess, the whole orchestra stands while the audience applauds. During intermission you can sit quietly, or get up to stretch your legs and talk with your friends. A bell rings a few minutes before intermission ends to warn you to return to your seat.

After intermission, you may hear a symphony. This long work is divided into four sections called "movements." There is a pause between movements, but the audience saves its applause for the end of the whole work. When the symphony is over, the orchestra once more stands to receive applause with the

conductor. The sound of clapping fills the hall. Then the conductor leaves the stage. At a signal from the concertmaster, the orchestra members file out. The lights go on, and the audience begins to leave.

All the way home, your mind keeps on hearing sounds and melodies from the concert.

You know that the minds of the orchestra members and conductor are also filled with thoughts of music. They are now planning practice and rehearsal time for the next concert. They will work together to make it as perfect a performance as the concert you have just heard.

GLOSSARY

audition—a test given to a musician who wants to join an orchestra

baton—the small stick used by a conductor to lead the orchestra

brass—the section of an orchestra in which the wind instruments are made of brass

concertmaster—the leader of the first violin section of an orchestra

conservatory—a special school for musicians

intermission—the short recess in the middle of a concert

movements—the divisions of a musical work

percussion—the section of the orchestra in which the instruments are played by striking or tapping

podium—the raised platform on which the conductor stands

program—the works of music on a concert. Also the booklet that lists them.

score—written music. In the conductor's score the parts for each kind of instrument appear separately on the page.

solo—music performed by one player

symphony—a long musical composition for orchestra. It has several movements.

tympanist—the leader of the percussion section and player of the kettledrums

woodwind—the section of wind instruments in the orchestra in which most instruments are made of wood